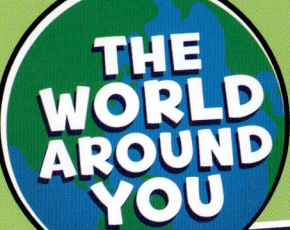

THE WORLD AROUND YOU

COUNTING AT THE SUPERMARKET

by
Christianne Jones

a Capstone company — publishers for children

Raintree is an imprint of Capstone Global Library Limited, a company incorporated in England and Wales having its registered office at 264 Banbury Road, Oxford, OX2 7DY – Registered company number: 6695582

www.raintree.co.uk
myorders@raintree.co.uk

Hardback edition text © Capstone Global Library Limited 2023
Paperback edition text © Capstone Global Library Limited 2024

The moral rights of the proprietor have been asserted. All rights reserved. No part of this publication may be reproduced in any form or by any means (including photocopying or storing it in any medium by electronic means and whether or not transiently or incidentally to some other use of this publication) without the written permission of the copyright owner, except in accordance with the provisions of the Copyright, Designs and Patents Act 1988 or under the terms of a licence issued by the Copyright Licensing Agency, 5th Floor, Shackleton House, 4 Battle Bridge Lane, London, SE1 2HX (www.cla.co.uk). Applications for the copyright owner's written permission should be addressed to the publisher.

ISBN 978 1 3982 4119 0 (hardback)
ISBN 978 1 3982 4120 6 (paperback)

Editorial Credits
Editor: Christianne Jones; Designer: Brann Garvey; Media Researcher: Svetlana Zhurkin; Production Specialist: Laura Manthe
Printed and bound in China

Image Credits
Shutterstock: aldegonde, (pears) 29, Alexander Dashewsky, (cucumbers) 29, Boudikka, 26, Caftor, 9, Denis Pepin, (peppers) 29, Dmitry Kalinovsky, 10, Emelie Lundman, 25, genkur, 14, Iakov Filimonov, top Cover, Icatnews, 17, Ivan Semenyuk, (avocado) 29, Juice Dash, 13, kaykhoon, 19, Kevin Khoo, 21, KK Tan, 7, Kwangmoozaa, 3, LightField Studios, 6, Lotus Images, (apples) 29, MaskaRad, (oranges) 29, Michael D Edwards, (watermelon) 29, OlegD, 16, Prostock-studio, 28, Roxane 134, 24, S1001, (lemons) 29, Shebeko, (carrots) 29, Sheila Fitzgerald, 8, 15, 20, Shutter B Photo, bottom Cover, Sorapop Udomsri, top right 26, StudioPortoSabbia, 11, Tricky_Shark, 22, Ttatty, 23, TY Lim, 18, vladm, 12, WNstock, (tomatoes) 29

British Library Cataloguing in Publication Data
A full catalogue record for this book is available from the British Library.

KEEP ON COUNTING

Items here. Items there.
Things to count are everywhere!
From the bottom shelf up to the top,
counting is fun at the grocery shop!

NAMING NUMBERS

1 one	**2** two	**3** three	**4** four
5 five	**6** six	**7** seven	**8** eight
9 nine	**10** ten	**11** eleven	**12** twelve
13 thirteen	**14** fourteen	**15** fifteen	**16** sixteen
17 seventeen	**18** eighteen	**19** nineteen	**20** twenty

COUNTING IN TENS

10 ten	**20** twenty
30 thirty	**40** forty
50 fifty	**60** sixty
70 seventy	**80** eighty
90 ninety	**100** one hundred

1
one

Let's look around for **ONE** sweet thing.
This pineapple will make your taste buds sing!

2
two

Find **TWO** vegetables that are not green.
These purple aubergines need to be seen!

3
three

Friday night is the big pasta dinner.
THREE jars of sauce will make it a winner.

4
four

A pile of oranges catches your eye.
You pick **FOUR** for your mum to buy.

5
five

These **FIVE** tomatoes are round, ripe and red.
They'll go with bacon and lettuce on bread.

6
six

It's time to decorate for Halloween.
These **SIX** plump pumpkins are the best we've seen!

7
seven

SEVEN ears of corn, yellow and bright, will make any meal a real delight!

8
eight

Cabbages aren't always green and round.
These **EIGHT** purple cabbages are easily found!

9
nine

The smell of fresh baguettes can't be beaten.
These **NINE** loaves will make your meal complete.

10
ten

Hearty soup warms you up in cold weather.
TEN tins are perfect for a big get-together.

11
eleven

Buy **ELEVEN** juice boxes for the Sunday brunch. Then mix all the flavours for a tasty punch!

12
twelve

Eggs are easy to scramble, to fry or to bake.
With a carton of **TWELVE**, what will you make?

13
thirteen

It can be smooth or crunchy and has a nutty taste, **THIRTEEN** jars of peanut butter will never go to waste!

14
fourteen

You need to buy snacks for a rumbling tummy.
Those **FOURTEEN** packs of crackers look so yummy!

15
fifteen

There are lots of hungry mouths to feed.
FIFTEEN packs of pasta are just what you need!

16
sixteen

Grandpa will cook his one special dish. He stocks up on **SIXTEEN** tins of tuna fish!

17
seventeen

SEVENTEEN iced doughnuts for after the play. A delicious dessert for the cast's big day!

18
eighteen

EIGHTEEN tubs of fruit brighten up the aisle. Pick your favourite fruit and make a small pile.

19
nineteen

Milk is refreshing and good for your bones, but you don't need to take **NINETEEN** cartons home.

20
twenty

TWENTY tubs of sweets are a dream come true. So many choices! Which one is for you?

There are **TEN** bananas in each bunch. Let's count in tens before we munch!

10 ten

20 twenty

30 thirty

40 forty

50 fifty

60 sixty

70 seventy

80 eighty

90 ninety

100 one hundred

COUNTING QUIZ

1. How many lemons are in the basket?
2. How many people are in the picture?
3. How many yellow peppers can you see?
4. How many baguettes are in the basket?
5. How many items are in the basket?

The answers are on page 30.

MORE OR FEWER QUIZ

1. Are there more cucumbers or more carrots?

2. Are there fewer apples or fewer oranges?

3. Is there an equal number of avocados and tomatoes?

4. Are there more watermelons or more peppers?

5. Are there fewer pears or fewer lemons?

The answers are on page 30.

COUNTING QUIZ ANSWERS

1. There are TWO lemons in the basket.
2. There is ONE person in the picture.
3. There are SEVEN yellow peppers in the picture.
4. There are TWO baguettes in the basket.
5. There are EIGHT items in the basket.

MORE OR FEWER QUIZ ANSWERS

1. There are **FIVE** cucumbers and **NINE** carrots. There are more carrots.

2. There are **SEVEN** apples and **THREE** oranges. There are fewer oranges.

3. There are **EIGHTEEN** avocados and **EIGHTEEN** tomatoes. They are equal.

4. There are **FIVE** watermelons and **FOUR** peppers. There are more watermelons.

5. There are **SIX** pears and **SEVEN** lemons. There are fewer pears.

LOOK FOR THE OTHER BOOKS IN THE WORLD AROUND YOU SERIES!

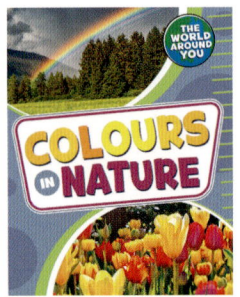  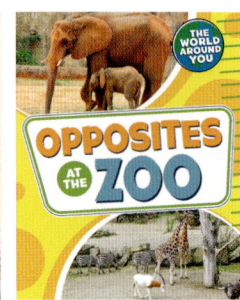 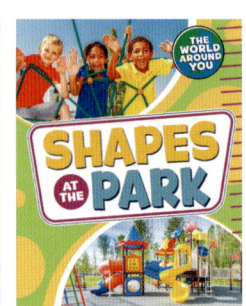 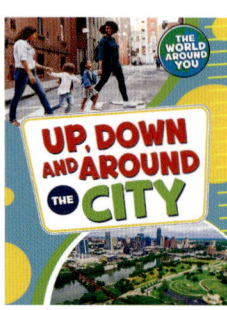

AUTHOR BIO

Christianne Jones has read about a bazillion books, written more than 70 and edited about 1,000. Christianne works as a book editor and lives in Minnesota, USA, with her husband and three daughters.